Krunchie s Cab

A Transport System for the 21st Century

Door-to-door automatic public transport

Krunchie Killeen

Contents

Introduction

When, in the year 2000, I began to re-visit my childhood idea of a tube-based transport system to replace motor-cars, buses and trains, there were no systems of this kind anywhere in the world. By now, however, many such systems have sprung up: automated pod-cars, called "Personal Rapid Transport" systems in multiple places, and, at Changzhou, in China, an operating system that resembles my proposal in many respects.

However, none of these new systems around the world quite get it together like Krunchie's Cab. Most emphasise the need to transport a lot of people at high speed. They miss the point that, when you go out your front door, your urgent need is not for speed, but for transport to get you directly from there to your local shop, school, restaurant, church, train station, theatre, friend's house, seaside, sports ground, public park, hospital, college or whatever. (The bus can't do that, because it has a fixed route, hitting some destinations only, so you still rely on a motor car for most).

Krunchie's Cab, however, can, indeed, take you directly to any of these places, in one non-stop journey, as well as to Paris or New York or whatever place, far or near, that you wish to visit. This is not just wishful thinking. The mechanisms for achieving this are described in this book.

All routes for Krunchie's Cab are combined in a network that joins local routes to each other and to inter-city routes, and national routes to international routes, with automatic routing to every destination.

Of course, the first implementations will be of single routes. A single route in Krunchie's Cab is multiple times more efficient than any competing bus-route. Soon after the first routes are implemented, they can be joined into a network and local networks joined into the larger network.

In my childhood, I visualised this "happening" without any help from me. It seemed really obvious that this "would happen," because that seems to be the natural progression , but the world is so slow at getting it all together, that I am forced to publish these clear instructions, for the benefit of all.

So, let me describe how it will work. You will walk out of your house and go no more than a hundred yards to a Cab-Stop. There you will find that there is already at least one Cab (capsule) waiting for you. You will show your Card (Travel Card, Credit Card or Mobile Phone) to a panel. This action will open the door of the Cab-Stop and the waiting Cab to you. You will select your destination: country, city, street and Stop (if there are multiple stops on that street) from a dedicated screen; and the Cab will take you directly to that destination, at 30 miles per hour; (and at faster speeds on longer routes).

Now, if you were taking a bus instead, you would not be making your journey at 30 mph. With all the stopping and starting at intermediate stops, and lining up with other traffic, you would be very lucky to make fifteen miles per hour.

Supposing you live five miles from town and the bus actually makes 20 mph, it would take 15 minutes for the bus to get into town. Add to that the indeterminate time you would spend waiting for the bus. Also, the bus has a fixed route and very often will not take you exactly where you want to go. You may have some walking to do, or have to get another bus.

Krunchie's Cab will take you that five miles in six minutes and always take you exactly to where you want to be.

This is not pie in the sky. It is perfectly feasible with today's technology. Moreover, it is utterly low-cost and multiple times more efficient than other forms of transport, and can have zero carbon foot-print and save the world.

When I began this project 23 years ago, I doubt if anybody was ready to listen. Now, however, many realise that the world faces disaster if we do no de-carbonise world transport, as Krunchie's Cab promises.

1. Description

A Tube-Travel System

Krunchie's Cab is a proposed transport system where Capsules travelling in Tubes provide door-to-door transit to members of the public. We call the capsules "Cabs."

You catch a Cab outside your door and it will take you automatically to your choice of destination anywhere in the network of routes, and potentially, anywhere in the world.

Capsule Size and Tube Diameter

A regular capsule would be similar in size to a motor car minus engine, bonnet, boot and wheels: approx. 4 foot wide by 6 foot long, comfortably seating 4 adults and their luggage. A *Cab* sits snugly in its Tube.

Door to Door Transit

Enter a Capsule at any street in any town in any country and exit at your choice of street in the same or any other town or country.

Automatic Routing

Key in your destination, (by selecting country/ town/ street/ stop), and the system delivers you straight to that destination, without any change of vehicle at any intermediate point.

No Driver

There is no driver. Access is by Mobile Phone, Credit Card or Travel Card; destination is keyed in, and the Cab automatically takes you to your destination.

Non-stop

Motion is continuous from starting point to destination. There is no stopping at intermediate points (except for the passenger's convenience).

Off-line Starting and Stopping

All stopping and starting is off-line (i.e., on sidings). As a result, the stopping and starting of one capsule does not hold up other capsules in transit.

24/7 service

There are no schedules or time-tables. A capsule can be engaged at any time of day or night.

A Network of Routes

Routes are inter-connected to form a Network, and Capsules are automatically routed through the Network.

Local routes, which, at c. 30 mph non-stop, are faster than current bus, tram and many commuter-rail routes, automatically connect to each other and to ring routes, inter-city routes and inter-national routes in which the capsules travel at much greater speeds.

Existing technology engaged

Existing pipe-manufacturers are able to produce the tubes.

Suitable Control Systems already exist that can automatically direct capsules through a network of routes.

Existing manufacturers are capable of producing, to specification, vehicles suitable to the system.

No Intersections, Traffic Jams or Accidents

Where routes cross, there is no intersection: one tube simply passes over the other. Capsules in one route can't obstruct capsules in others. There are no traffic jams or collisions, and no pedestrian collisions are possible.

Single Circuits

Of course, a system will often commence as a single circuit. This will carry about a hundred times more passengers than a bus route operating on the same line, will never keep them waiting, and carry them swiftly to their destination, with no intermediate stops or hold-ups en route.

How It Is Done

How can all this be achieved?

The essential elements that make *Krunchie's Cab* possible are discussed in Chapter 6 (*"Vital Concepts"*) below.

Naming it Krunchie's Cab

"Krunchie" is my nick-name; "Cab" is short for "a Capsule that takes you door-to-door like a taxi;" so I call the system "Krunchie's Cab."

2. Genesis

I learn of Obsolescence

When I was four years' old, I asked my mother if I could have a ride in a stage-coach. She told me that stage-coaches no longer existed: that they were made obsolete by the steam-engine; but that now trains were being made obsolete by the motor-car.

This conversation took place in 1947. My mother's predictions were correct. By 1965, all Ireland's tram systems had been dismantled and over half its railways closed down, all replaced by road transport.

I wonder what would follow Motor-cars

In the months and years that followed that conversation, as I played out transport systems in our back yard, I wondered what mode of transport would make the motor-car obsolete.

Back garden transport system

The Solution – Tube Travel

I figured that such a system of transport should, like motor-cars, take users Door-to-Door, and at their own choice of time. The system I visualised that could do this was a system of Capsules travelling in Tubes.

The Webbed Network

At first this system appeared complex, with a tangle of tubes encircling the globe and visiting every house-hold. However, I worked out how the tangle of tubes could be unravelled –

- a web of local circuits, like a net of chicken-wire, reaching every local destination, and

- straight tubes, fed by the bendy local tubes, connecting cities and continents.

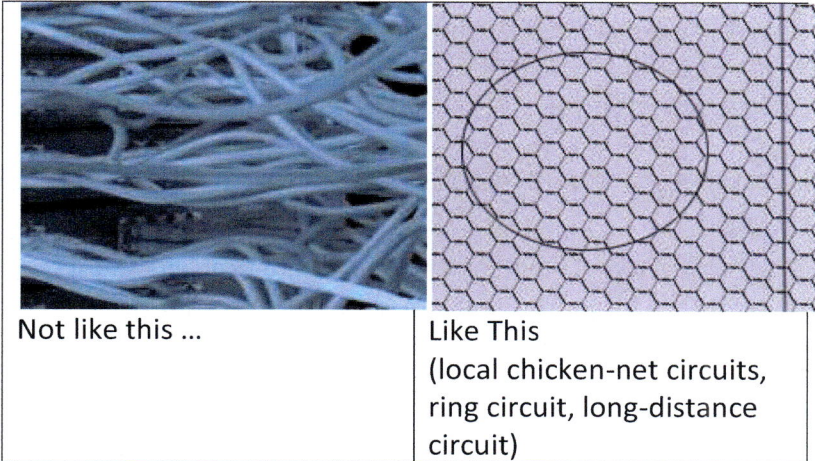

| Not like this … | Like This (local chicken-net circuits, ring circuit, long-distance circuit) |

Of course, the local circuits would not rub shoulder-to-shoulder like actual chicken wire, but be connected by connecting tubes as follows:

11

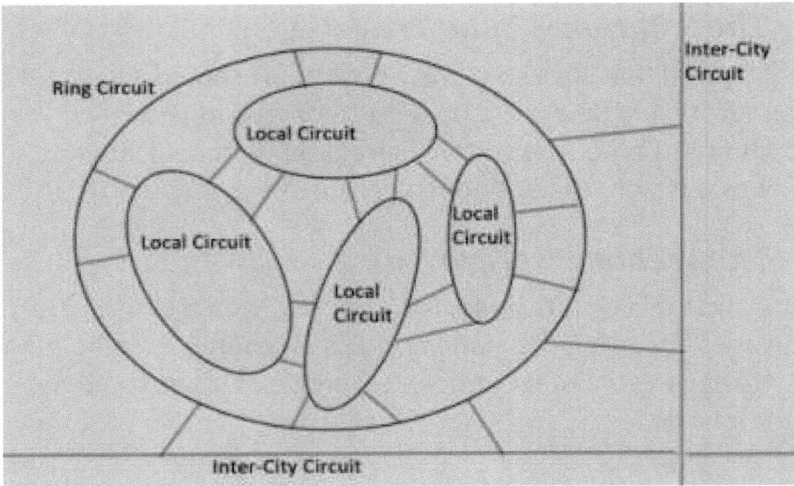

Houses already connected by tubes!

When my father cleared a blockage in the local sewer, I knew my plan was feasible. I observed that the houses were already connected by a tube: the sewer pipe. The narrow pipe, serving the houses, emptied into a wider district pipe. A little wider, and it could accommodate passenger-capsules.

Passenger-tubes easier to install than sewer-pipes

Passenger-tubes did not need to be underground and would take up no more ground-space than a footpath. They did not need to have an outlet to every house: one capsule-stop could serve a group of houses.

The Circuit

Of course, the sewer pipe took its contents in one direction only. A passenger tube would circle back to form a circuit on which, though still one-way, every premises would be accessible from any starting-point. Vehicles would neither stop nor start in the circuit, but on sidings at destinations.

Low Friction and Air-resistance

In a tube free of friction and air-resistance, a capsule, projected into the circuit, would continue in perpetual motion without any further propulsion. In a real system, however, propulsion would be boosted en route.

Cabs to Transmit their Destination

In my childhood version, to route the vehicles to their destination, each capsule would carry a sign showing its destination, and shunt levers would be set by workmen. Revisiting the scheme after 50 years finds the workmen replaced by computers and the routing automated. I realised that the same basic algorithm used by telecommunications systems to route signals across the world could be used to route my Cabs.

The Time is Now!

To me as a child, all this was for the distant future. The motor car would first have its day. Now I find that the motor car's day is done.

In the year 2000, I read an account of a report by Ireland's *National Transport Authority* that road-building to cater for the ever-increasing number of cars was unsustainable (and other forms of transport should be developed).

I suddenly realised that the days of the car were numbered and my transport system had become relevant. The distant future had already arrived and Krunchie's Cab was needed now!

3. From Project to Pamphlet

Patent Advice

When I realised the time had come for my childhood transport plan, I sought professional advice on applying for a patent. The negative response had three main points:

1. An outline description, like mine, was not sufficient; there must be a detailed technical description, such that a person or team with skills in the area could construct the patented object or system from that description alone.

2. A premature patent was inadvisable. One should not apply for a patent until ready to go into production: otherwise the patent, as often happens, might have run out even before the project got off the ground.

3. Novelty must be proved. The "idea" of capsules travelling in tubes is as old as the hills. There is nothing new in the idea of a Circuit. Sidings are already there. It is the combination that is novel, and the Patent Office might not be easily convinced that this is a technical innovation.

"Commercialising your Research"

I then did a course in "Commercialising your Research."

My business plan at the time was:

1. An Outline Description of the Project (done)

2. Phase 1 Venture Capital (with business advice)

3. Establish a Limited Company

4. Engage a skilled team to produce a Desk-Top model

5. Phase 2 Venture Capital

6. Engage with a Land Owner (e.g., City Council, Railway Company, Canal Company, Board of Works, or large private firm) to produce a life-size Prototype.

7. Promote and Licence the project to actual users.

Commitment

It was apparent that the Business Plan would not take off without commitment of time and resources from myself. A Venture Capitalist would want to see evidence of this commitment first.

The Soft Option

I lacked commitment: I was not going to re-mortgage my house, abandon my family, or wear sack-cloth and ashes for the sake of the project.

I was more of an arm-chair philosopher, and opted just to bring out a booklet describing the project; and let whoever might be interested take it up.

This project had to stand in line with my other interests, all described in my blog at *https://krunchiekilleen.blogspot.com*.

It was 2017 before the first edition of *Krunchie's Cab* came out, under the title *Transport 21 Hundred*. The present volume is an update of that (much having happened in the last few years).

My Chinese Competitor

Initially my project was totally novel. However, it is said that inventions often happen at the same time at different points on the globe. While I was wondering what to do about my invention, Mr Nanzheng Yang MBA was assembling a well-funded team in China for the same purpose.

A Flurry of Activity

Over the last 20 years, and accelerating, there has been a flurry of projects around the world devoted to developing alternative transport to the motor car. However, all these (except *Krunchie's Cab*) miss the essential element of catering for the multiple local destinations.

4. Previous Systems

Metro Systems

When rail systems for metropolitan areas use tunnels, they are often referred to as Tubes. However, they are really not Tube systems but railway systems.

London Underground transport system (Hammersmith Station)

To build a railway or tunnel is a massive and expensive project. To construct a *Krunchie's Cab Circuit* is simple and inexpensive. Simply lay down the tubes on the ground! *Krunchie's* Tubes require little site preparation and can be laid on any surface, wet or dry.

The Ancient World

The idea of human transport via capsules travelling through tubes is mentioned in the literature of ancient China and Greece, but never actualised in those times. Pipes carrying water or sewerage were quite common, as were railways, i.e., systems where carriages were pulled along rails. The carriages were pulled by animals or humans, over ground or through tunnels, or propelled by gravity and hoisted back by manually-operated cables.

Pneumatic Propulsion

Invented by William Murdock around 1799, small parcels, including telegrams and letters, were propelled in capsules through pipelines by means of compressed air and a partial vacuum. Postal services in many cities used such systems from the middle of the 19th century until the end of the 20th.

New York pneumatic post (picture by Bain News Service, 1914, via Wikimedia LLCCN2014697406.tif)

A major network of tubes in Paris ("The Paris pneumatic post") was in use up to 1984, when it was abandoned in favour of Fax. The *Prague Pneumatic Post* continued up to 2003 (Wikipedia).

Pneumatic propulsion was also popular in department stores, for transmitting cash and invoices between office and sales posts. It is still used in hospitals and laboratories for transmitting blood samples for analysis, and in many factories for transmitting small parts.

 Krunchie's Cab does not propose to use compressed air, but rather to move capsules by means of electromagnets.

Dalkey Atmospheric Railway: 1844 – 1854

A tube was laid between railway tracks (Dalkey to Dun Laoghaire, Ireland). A piston, suspended from a train ran through the tube through a sealable slot in the top of the tube. This piston, projected by compressed air, carried the train.

The Dalkey Atmospheric Railway (Illustrated London News, Public domain, via Wikimedia Commons)

Pneumatic Metro Systems

When Metro Systems were first proposed in the latter half of the 19th century, one option considered was transport by means of capsules propelled through tubes by Pneumatic Propulsion.

The Pneumatic Passenger Railway, as Erected at the American Institute, Fourteenth Street, New-York, 1867 (picture by Alfred Ely Beach, via Wikimedia Commons)

The drawbacks were:

- Only one capsule could occupy a tube at one time;
- Propulsion was from point to point, not allowing intermediate stations;
- Motion started and stopped with quite a jerk.

While tubes were cheaper than tunnels or rails, (though not as cheap and convenient as now), propulsion by compressed air would be expensive.

Krunchie's Cab has none of these drawbacks, and provides a cheap, efficient service, where tubes can accommodate many capsules at one time, and deliver passengers automatically to any destination attached to the network.

At the time, rail systems were seen as more versatile and practical than Tubes. This is no longer the case.

Maglev

Magnetic Levitation, or *Maglev*, uses two sets of electro-magnets. One is used to lift a vehicle a few centimetres off the ground, eliminating surface friction, and the second set propels it forward. It entails a magnetised metal track under the vehicle, to raise it, and another one or two tracks, often above or beside the vehicle, to drive it forward.

Many such systems have been closed down after a fairly short time, because of the expense of operation, but there are several successful systems in use today (for example in Japan, China, Germany and Switzerland).

Maglev rail systems still hold the record for land speed despite the recent arrival of the theoretically faster Super-High-Speed *Hyperloop*.

Japanese Bullet Train (Photo, 2005, by "Josemite" via Wikimedia Commons)

In Japan, *maglev* systems have been greatly expanded over the last 20 years. This has not solved the problem of road congestion, because *Maglev Railways* do not deliver doorstep to doorstep, as motor cars promised, and as

Krunchie's Cab achieves, but drop passengers at hubs. Passengers still need motor cars to get home, so congestion continues to worsen. So, step in *Krunchie's Cab* to take you to your doorstep!

Hyperloop

Robert Goddard, in 1910, proposed a train, elevated by *maglev*, but running in a low-pressure tunnel, that would travel from Boston to New York in 12 minutes (at 1,000 mph). Elon Musk revived the idea in 2012 and coined the term *Hyperloop*.

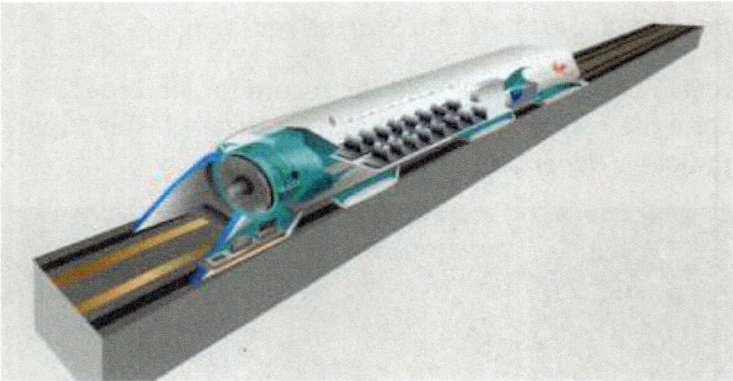

Hyperloop concept (image by Camilo Sanchez, 2015, via Wikimedia Creative Commons): a capsule, called a "Pod," travelling in a Tube, with Maglev propulsion

Since then, several companies have pursued the idea. *Virgin Hyperloop* appears to be the first to market:

"On November 8, 2020, the first passengers travelled safely on a Hyperloop – making transportation history" (https://www.virginhyperloop.com)

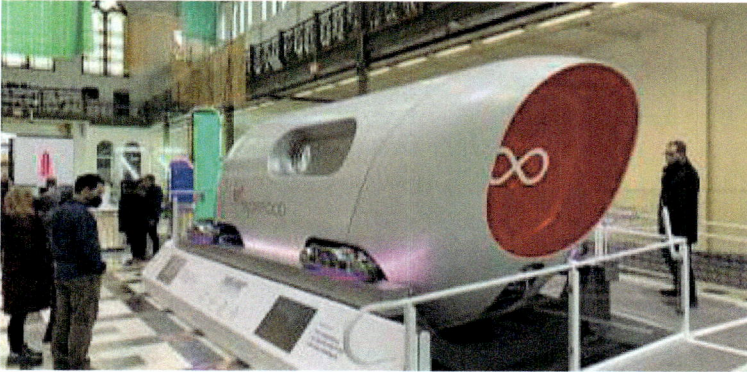

Virgin Hyperloop XP-2, also known as "Pegasus," (picture by APK, 2022, via Wikimedia Creative Commons)

The *Virgin Hyperloop XP-2 pod* is shorter than the original *Hyperloop pod*, and is coming closer to *Krunchie's Cab*, but still way, way more complex and expensive and impractical.

While the maximum speed so far reached by *Hyperloops* is 288 mph, (at a competition hosted by *SpaceX* in California in 2019), the *Virgin Hyperloop* launch achieved 107 mph. The passengers were apparently uncomfortable with the sudden acceleration to this speed. Elon Musk tweeted that he would produce a *Hyperloop* that would be driven by the energy of passengers' screams.

All this supports the approach of *Krunchie's Cab*, which aims to accelerate to no more than 30 mph in the first few seconds, and builds up larger speed step by step on longer routes. Like previous *Maglev* systems, *Hyperloop* delivers hub to hub, and needs *Krunchie's Cab* for the completion of the journey to your doorstep.

Transoceanic systems

Under-sea pipelines are common for delivering oil and gas. *Krunchie's Cab* can use oceanic pipe-laying technology to provide under-sea routes. This is enormously cheaper than constructing under-sea tunnels.

Freight Tube Transport in Germany

CargoCap.com

To avoid delays by traffic congestion in roads, some German cities have an underground tube system for carrying goods (see "CargoCap.com"). These tubes are approximately 2 feet in diameter. The underground tunnels are developed by tunnel-boring machines. This system is now proposed for Abu Dhabi and Dubai.

Krunchie's Cab tubes would be at road level, (except where they can conveniently be accommodated by existing tunnels or raised on stilts), would have a diameter of about 4 feet, enable human transport, and automatic routing to any destination attached to a network of routes.

Personal Rapid Transit

Over the last 15 years, and accelerating, systems called *Personal Rapid Transit* have sprung up at many places around the globe. These use *maglev* and have automated pods that travel on rail, usually elevated on stilts, and have off-line start and stop like *Krunchie's Cab*. Pods are designed, like trains, to allow passengers to stand up. This is unnecessary; people do not stand up in motor cars. *Krunchie's* capsules are sit-in, sleeker and more aerodynamic.

Skyweb Express pod and track	Off-line station in Washington State

These PRT pods, carried on rails rather than in tubes, are a lot more expensive than *Krunchie's Cab*. However, they are less versatile and less capable of being expanded into a wider, worldwide network. Instead of this elongated offline station in Wahington State, (pictured here), *Krunchie's Cab* would provide two adjacent Cab-Stops, reducing the need to queue. You should always find a Cab waiting for you when you arrive at a Cab-Stop.

There are about 20 proposed PRT systems in the world awaiting finance at the moment. This finance would advisably be shifted to *Krunchie's Cab* systems.

Chinese Tube-Net Transit

In 2001, just as I was re-opening my childhood vision of tube travel, Mr Nanzheng Yang DBA was assembling a research team called *Tubenet Transit Institute* for the same purpose. He obtained Patents in China and the USA and other countries for:

> "... a new personal transport system for passengers and goods, which comprises a self-powered closed tube car carrying passengers or goods from door to door; and a network of tubes comprising a plurality of tubes."
> (**USA Patent number:** 8006625).

He launched a pilot version of his system in Hangzhou City in China in 2018, and aims to implement it around the world.

A driverless tube-travel system, at first sight just like *Krunchie's Cab*, there are multiple differences:

- *Tube-Net* uses *Maglev*. There are metal tracks beneath and above the pods to elevate and propel them and guide them on their way.

- *Tube-Net Pods* journey "from building to building." It is not clear how local and main routes interlink.

- All *Tube-Net* routes are raised on stilts, whereas *Krunchie's* are laid over wet or dry ground or under water, or raised on stilts, as circumstances require.

- The *Tube-Net Pod* is upright and walk-in, like a carriage, whereas *Krunchie's Cab* is sit-in and sleek like a motor-car.

- *Krunchie's Cab* fits like a glove into its Tube. The *Tube-Net* pod travels in a tunnel-like space surrounded by air.

- *Tube-Net Pods* are designed to travel at 40, 60 and 80 kph. *Krunchie's* start at much the same for local routes, but rise to much larger speeds for long distance routes.

Tube-Net absorbs solar power, which it feeds to batteries, which drive the pods; but it uses a connection to the electricity grid as backup. It claims to provide transport at 1% of the cost of motor traffic. It is said that a tube can carry up to 34,000 passengers per hour (or over 800,000 per day).

Nevertheless, *Krunchie's Cab* will be better and more versatile, cheaper; and easier to construct, operate and expand to a world-wide network.

5. Distinction from other Systems

A Different Objective

I come to Tube Travel from a different angle than other proposed or implemented systems. My objective is to replace motor-cars and buses, and eventually airplanes. To replace cars, you need door-to-door transit. A user will catch a Capsule outside his front door that will take him all the way to his destination, wherever that may be.

The objective of other systems is to increase the speed and capacity of routes between large hubs. The public still has to make their way home.

If I want to go to some place that is not on a direct bus-route from my home, whether it be a school, a hospital, a shop, a concert hall, a restaurant, a historic site, a football ground, or a cousin's house, I am not interested in Super High Speed; what I want is direct transit to that place. *Krunchie's Cab* is the system that will deliver this.

When I want to travel a greater distance, I can expect *Krunchie's Cab* to build up speed on the way.

Propulsion Options

I tentatively propose electro-magnetic propulsion, but several other means of propulsion would be feasible. The brand new Tube-Net system in Hangzhou city in China (mentioned above) uses maglev. Bear in mind that cost comes into the equation. Magnetic Levitation (Maglev) would possibly enable greater acceleration and speed, but at much greater cost. Maglev from skyscraper to skyscraper might be affordable, but not to every citizen's residence.

Door to Door Transit

Krunchie's Cab provides Door-to-Door transit, instead of transit from Station-to-Station like previous systems.

Distinction from Personal Rapid Transit (PRT)

Personal Rapid Transit systems can have automatic routing, but operate on tracks rather than in tubes, and are less flexible, more expensive to develop, and less capable of expansion into inter-city and international networks. Nor have their developers given attention to the *"Webbed Network"* I described in *Chapter 2* (and re-visit in Chapter 6) to facilitate the spread of the system all over the globe.

Distinction from Hyperloop

Hyperloop systems are expensive to implement, will bring passengers from Hub to Hub, rather than from door to door, and only suitable for connections between large centres of population. They take up a vast area of ground space and require straight paths between destinations. They leave passengers to arrange their journey from Hub to Home, a journey *Krunchie's Cab* can take on.

Krunchie's Cab, proposed mainly to deliver door-to-door transit, is the only proposed system that can defeat the congestion caused by motor cars. To replace motor cars, you must provide a public system that delivers door to door transport.

Krunchie's Cab routes can be arranged to take passengers through progressively faster *Circuits*, achieving Super High Speed eventually if the Route is long enough, (and back down step by step, before halting at your doorstep).

Distinction of Cost

Krunchie's Cab is a cheap transport system, easily and quickly implemented, requiring no site-preparation, suitable for local transport as well as inter-city and international travel, and capable of rapid expansion by using existing road, rail and canal routes.

Tube travel will cost about 1% of comparable road travel (per the Chinese experience with *Tube-Net Transit System*).

Small Space

Krunchie's Cab takes up no more space than a footpath – even less space than one lane of motor traffic. It can also be elevated on stilts, thereby taking up no road space at all. One tube can accommodate the equivalent of four motorway lanes.

Hyperloop is said to require a straight passage seven metres wide which needs to be cleared of buildings and obstructions for construction to proceed.

Built-in Super-speed Circuits

Starting at your doorstep in a low-speed local Circuit, a Route can upgrade speed-wise to Super Fast Speed in long-distance routes.

6. Vital Concepts

What makes Krunchie's Cab work?

How does *Krunchie's Cab* deliver a passenger, without stopping at all en route, from any starting point (in a network of routes) to any destination, automatically finding the best route to get him there?

In this chapter, the vital concepts that achieve this are described.

Circuits

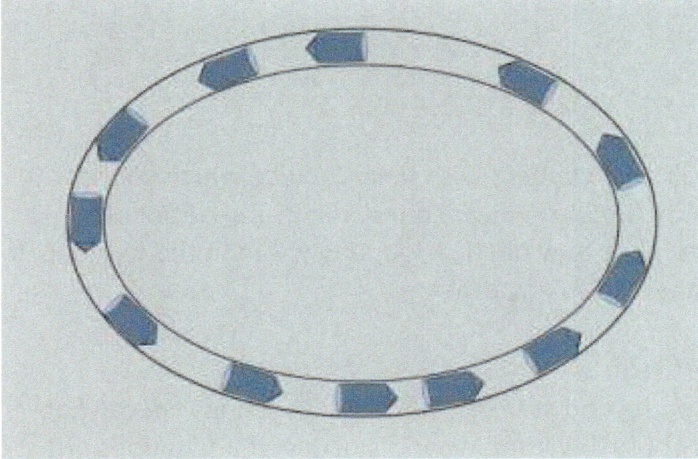

Cabs travel in Circuits (not necessarily circular or of regular shape). We reserve the word "Route" for the path between a Starting Point and a Destination. A Route often traverses several connected Circuits.

All Cabs in a Circuit travel in the same direction and at the same speed, more or less. (Factors such as gradients can cause a fluctuation in the actual speed, slowing as a cab climbs a hill and accelerating coming down). There is no stopping or starting within a Circuit.

A Network of Routes

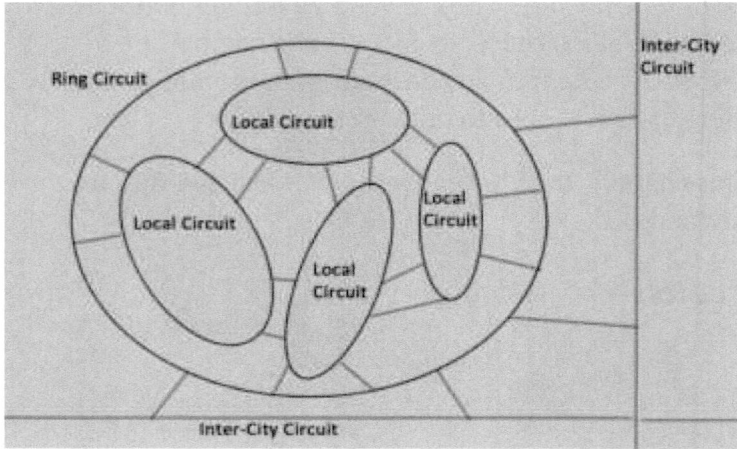

Local Circuits connect with Ring Circuits, which connect to Inter-City Circuits, which connect to Trans-National Circuits. Thus, any destination, anywhere in the world, can be connected to any other.

Cab Stops

All stopping and starting is off-line, in sidings we call Cab-Stops. Thus, the stopping and starting of a Cab does not inhibit the movement of other cabs in the Circuit.

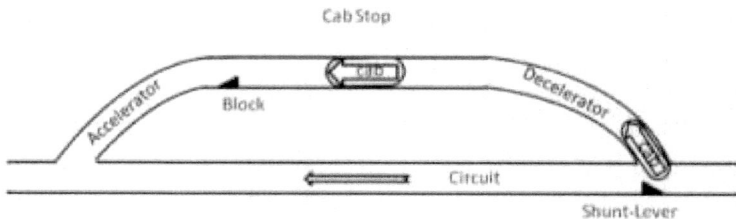

A shunting device (or "Shunt-lever") is activated automatically to move a Cab from the Circuit to a Cab-Stop. The Cab then enters a Tube, called a Decelerator, where its speed is reduced, and it is brought to a stop.

A "Block" is set to prevent the Cab from starting again unless the Exit to the Circuit is clear.

When the Cab starts, it goes into an Accelerator, which is a Tube that brings its speed up to that specified for the particular Circuit.

Several Cabs can wait for customers at a Cab Stop.

Connectors

Every Circuit will have a specified Speed. The Network may contain Circuits of differing speeds, e.g., local circuits of low speed and urban or rural Ring Circuits of a higher speed, the latter connected to inter-city Circuits at a still higher speed.

Circuits are connected by Connectors, Accelerators and Decelerators.

Connection between two Circuits of similar speed is made by Connectors. These are Tubes which deliver Cabs to the receiving Circuit at the same speed as they left the donor Circuit.

Connection between two Circuits of differing speeds is made by Accelerators and Decelerators, which increase or decrease the Cab's speed as required.

Route-Maps

Identifiers

Every Cab, Circuit, Cab-stop and Shunt-lever/ Junction will have a unique identifying number.

Computing the Route-Map

When a user keys in his destination, the Starting Point and Destination are communicated to the Network Controller (a computer). The Network Controller computes the Route from Starting Point to Destination, and communicates the Route-Map to the Cab. In doing so, it takes account of the projected traffic throughout the Network.

The Route-Map

A Route-Map consists of the identities of the Starting Point and Destination, as well as a list, in sequence, of the identifiers of every Junction where the Cab is to switch from one Tube to another.

Switching Circuits

As a Cab travels on its way, it transmits its own identity and the identity of the next junction on its Route. When it approaches the specified Junction, the message causes the Shunt-lever to switch, thereby redirecting the Cab to its new Circuit or Stop. After the Cab has passed through, the Shunt-lever automatically resets, and the Cab starts transmitting the identity of its next relevant Junction.

Speed-Monitoring and Control

As a Cab travels through the network, it constantly transmits its Identity. Sensors along the Route, receiving the Signal from the Cab, match the Cab's location to time and place, which is automatically communicated to the Network Controller.

The Network Controller constantly calculates the speed of the Cab from this information and applies a speed boost or brake as required to maintain the Circuit Speed.

Collision Avoidance

A Cab-stop will inhibit a parked Cab from starting while another Cab in transit passes the Stop.

The Network Controller will slow Cabs in Interconnecting Tubes while another Cab passes the Junction.

7. The Obsolescence of the Motor-car

The promise of the motor-car: travel to any destination at any time, day or night:

Photo by Roger Kidd, 2010, via Wikimedia Creative Commons

The reality today: gridlock; destinations inaccessible; unsustainable road development and maintenance:

Photo by Sergey Norin, Moscow, 2011, via Wikimedia Creative Commons,

In the next picture, below, I show a back-seat passenger's photo of a trip on London's M25 motorway, October 2022. Due to some obstruction on the motorway (the likes of which apparently occurs almost every day), one lane is closed and speed on the other lanes is reduced to 40 mph at this moment, (but also came to a complete stop once or twice during the journey).

England has invested billions in developing its high-speed railways, but this does not solve the problem of congestion of roads and motorways, because trains only take passengers to stations and not to their ultimate destinations. Motor-cars are still required to take people home.

This particular trip by motorway was necessary because there was no bus, rail, or underground connection between our accommodation in Hammersmith and our destination in Seven Oaks, Kent. This trip of 60 miles (via M25) should have taken 1 hour and 25 minutes, by motorway, but because of congestion, took us over 3 hours.

According to *Wikipedia*:

> There are 4 ways to get from Hammersmith to Sevenoaks (Station), i.e., by subway, train, taxi or car

RECOMMENDED OPTION

Subway, train via London Charing Cross
Take the subway from Hammersmith to Embankment Take the train from London Charing Cross to Sevenoaks
1h 19m €15 - €30

3 ALTERNATIVE OPTIONS
Subway, train
Take the subway from Hammersmith to Blackfriars station
Take the train from London Blackfriars to Sevenoaks
1h 51m €17 - €21

Taxi
Take a taxi from Hammersmith to Sevenoaks (Station)
53 min   €150 - €190

Drive
Drive from Hammersmith to Sevenoaks (Station)
53 min   €11 - €17

By *Krunchie's Cab*, all four motorway lanes here could be replaced by one tube; the speed would be c. 200 mph, and we would have arrived at our destination in less than half an hour. There would never be congestion on the route. There would be no need for a taxi to and from the train stations, and there would be no need to change stations en route. *Krunchie's Cab* would take you from your accommodation in Hammersmith directly to your destination in Seven Oaks (or wherever).

If operating normally, the Motorway offers a slightly faster route (between stations) than the Underground system. This is because the underground entails some waiting, and trains stop at intermediate stations, and one changes line en route. With *Krunchie's Cab* there would be no waiting, no stopping en route, higher speeds, no line-switching, and door to door transit.

8. Current Capital Wastage

The motor-car system is replete with wasted resources. At any time, over 95% of the world's cars are parked or otherwise not in service. (In Krunchie's Cab system all capsules are moving or actively waiting for passengers at all times).

Parking Lot by Dean Hochman, 2014 (16161860403.jpg, Wikimedia Creative Commons)

Railway systems absorb an enormous amount of land:

Railway land in Dublin, Ireland

Large stretches of land can be reclaimed for amenity and development when railways are replaced by Krunchie's Cab.

Road-construction absorbs a ridiculous amount of capital -- redundant with Krunchie's Cab:

A spaghetti junction by Highways Agency on flickr (via Wikimedia Creative Commons)

9. Present and Future Rush-hour

Present Rush-hour traffic at Parnell Square in Dublin

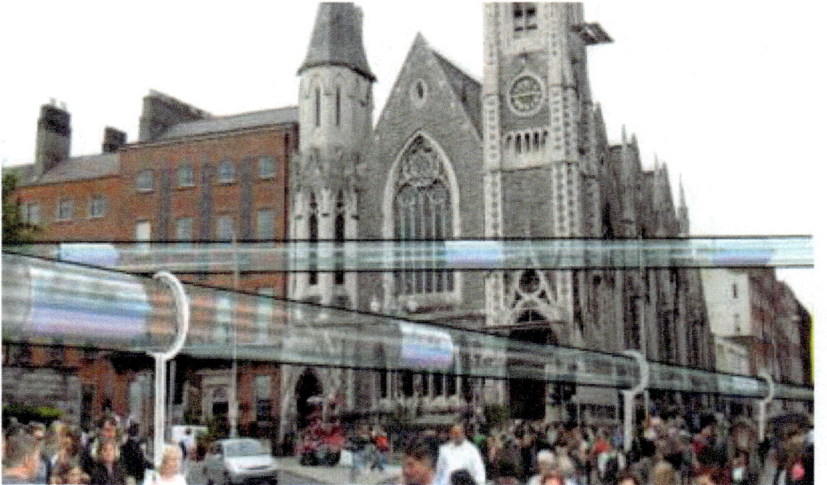

Future Rush-hour traffic at Parnell Square (artist's impression; obsolescent vehicle included for comparison):

10. Network Schema

Bendy routes (coloured green in the illustration) weave their way through localities, reaching close to every doorstep and feeding into straighter tubes (coloured red) that provide greater speed between urban centres.

The straighter the tube, the faster a Capsule can travel. Each circuit would have a specified speed, bendy circuits being slower than straight ones.

Local Circuits at c. 30 mph,

> Feed into Urban or Rural Ring Circuits at 100 to 200 mph, which

> Feed into Inter-City Circuits at 400 mph, which

> Feed into Trans-National and Trans-Continental Circuits at 800 mph or greater and ultimately into Super High Speed Systems at up to 3000 mph.

11. Dublin City Example

Ring Routes

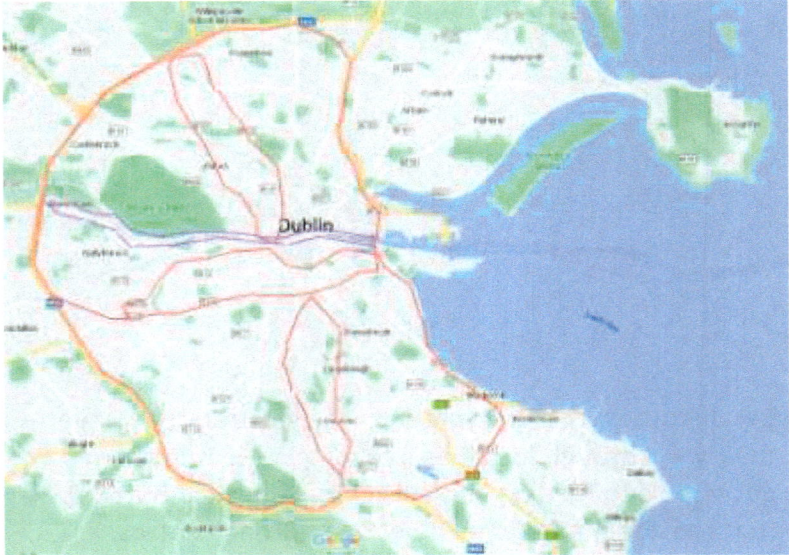

Here we show possible main or "Ring" circuits of *Krunchie's Cab* for Dublin City:

An Outer Ring Circuit, coloured red, which follows the M50 (coloured yellow) most of the way.

An Inner Ring Circuit, a Northern Ring Circuit, and a Southern Ring Circuit, also coloured red.

Liffey-side Ring Circuit coloured purple.

Local Circuits would generally follow the pattern of existing roads and would feed into these Ring Circuits.

A speed of 200 mph would be suitable for the Outer Ring Route, and speeds of 100 mph for the others. All the Ring

Routes would be interconnected (and connect to Inter-City Circuits not shown in this diagram). It would take no more than half an hour to complete a journey to any city destination.

A Ring Circuit, of course, would deliver you back to a local Circuit, which would deliver you to the very door-step of your destination.

Routing would be automatic. No matter what point you start at, you select your destination (From Street Name and Stop Number), and Krunchie's Cab automatically takes you there.

Liffey-side circuit

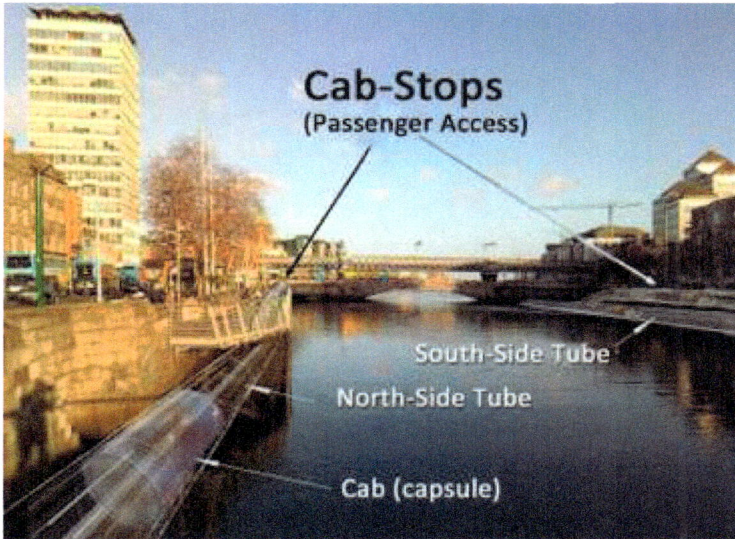

Riverside Tubes, Dublin

Here, we show a *Krunchie's Cab* Local Circuit, anchored to the walls of the river Liffey in Dublin. This circuit, going up one side of the Liffey and back the other side, could handle

500,000 passengers a day and, of itself alone, enable the pedestrianisation of the Liffeyside roads, (or "Quays"), and free the river-side restaurants to serve food *al fresco*.

This local circuit could feed into the faster Ring Circuits in the previous picture for people going on longer journeys at greater speed, and, of course, into a network of routes for delivery of passengers to any destination in the world.

12. Connecting Rural and Remote Communities

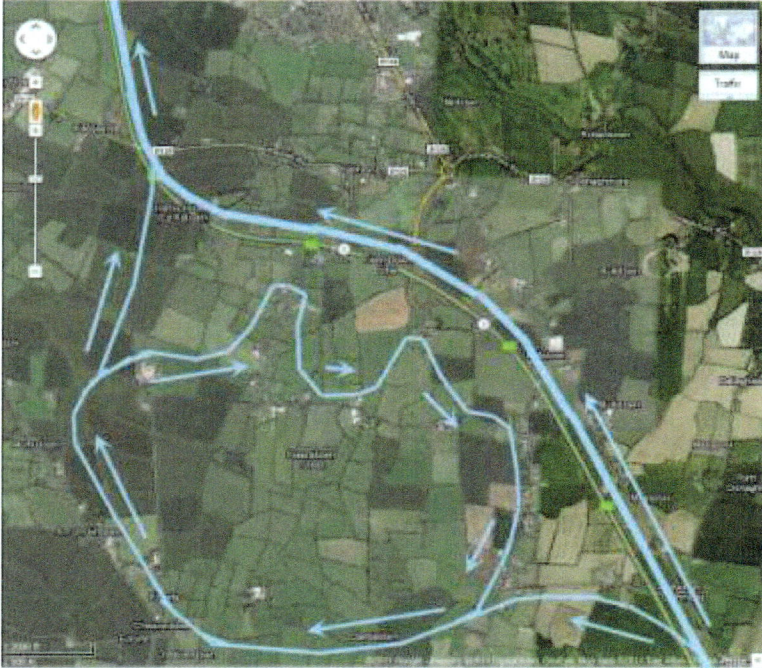

Rural and remote communities might not provide sufficient traffic to make a local route profitable. However, members of such communities would benefit significantly by being connected. The solution is to allow such communities to connect to the network on payment of a fee to cover the connection cost, just as new users pay a fee for connection to the electricity grid.

For a once-off capital investment of about the cost of a motor-car, connection could be achieved, after which all the benefits of the network would be available. Unlike the investment in a motor-car, which must be renewed every few years, this connection fee would be once off.

The image shows a bendy route, winding its way through a rural community, connected to a rural ring-circuit, (the latter shown wider for the purpose of illustration only).

No residence, no matter how remote, needs to be excluded from the system. Country towns would benefit enormously, since people could travel there at their choice of time, rather than being limited to schedules. Krunchie's Cab would be a life-line to rural communities.

13. Going Places

Access to Tourist Sites – only by Car

One day, I fancied visiting Boora Bog, a nature reserve and eco-tourism site in the middle of Ireland. I looked up the Web to see if I could get public transport to the site, but no! The only access was by motor-car. (I could get a train or bus to the nearest urban centre, but require a taxi from there). This is but one of thousands of heritage and tourist sites in Ireland effectively only accessible by car.

Access to Hospital for Elderly

Elderly people in Glasnevin, the area of Dublin where I live, are very well served by the *Healthy Ageing Clinic* of St Mary's Hospital. During the Celtic Tiger, the Eastern Health Board provided them with transport to St Mary's. The austerity measures that followed, however, saw an end to this free transport. With no direct public transport route, I witnessed several elderly foregoing their appointments in the hospital because of the cost of hiring a taxi to get there. Just one example of services made inaccessible or difficult by problems inherent in the current transport systems!

Car-Pooling

When my children were of school-going age, there was no direct bus route to their schools, and car-pooling was the only reliable travel-plan. Another example!

Krunchie's Cab solves the problems

When Krunchie's Cab is widely implemented, all these problems will belong to the past. Every destination will be

Krunchie's Cab

accessible in a single, on-demand, low-cost trip, from outside your door right to your destination, wherever it is.

14. Service Stations and Capsule Storage

A *Service Station* is a facility that could be attached to a tube *Route* to provide services such as food, first-aid, rest, repair and maintenance, security, or the storage of spare *Capsules*. Such facilities could include:

- *Food-Stations* providing restaurant or café facilities,

- *Repair-Stations* where *Capsules* would be directed for maintenance and repair,

- *Emergency-Stations* where *Capsules* would be directed when *Passengers* require medical assistance,

- *Security-Stations* controlled by security personnel,

- *Goods-delivery Stations* where users could pick up goods sent by Cab;

- *Rest-Stations* where *Passengers* could avail of hotel or hostel facilities, and

- *Cab-Stations* which would hold a store of *Capsules* available to be directed to *Routes* and *Cab-stops* which are short of *Capsules.*

15. The Travel Card

Automated Payment System

Payment for service would be by an automated system. This could be achieved by Credit Card or by Mobile Phone, but I confine discussion here to one such method, viz., the "Travel Card."

Pre-paid Card

A Travel Card would be a card which would "contain" a prepaid sum from which the cost of a proposed journey would be automatically deducted.

Present Card and Key-in Destination

A user would present the Travel Card to a Card Reader at the Cab-stop, and, once he has indicated his destination (by choosing Country/ City/ Street and Stop) the cost of the journey would be deducted from the balance "contained" in the card.

Travel Card as Identifier

A Travel Card would, also, be an Identity Card. The system would keep a record of journeys made and be able to relate any previous miss-conduct to the particular user. On applying for a Travel Card, the user would have his photo taken and this would be stored in the system.

Facilitate Passport control

In relation to international travel, the Travel Card would be able to record details of the user's passport and visa to facilitate border-crossing. In due course, arrangements would be made with the civil authorities of various countries to automate border controls. In the meantime, where automated processes are not in place, or are suspended, international travel could involve stopping at a Station for border and customs control.

Crowd Financing

Travel Cards could be used as a means of Crowd Financing. Where a Card is in credit, the balance could be used to fund the expansion of the network. The public could be invited to place fairly substantial sums on their cards, which would attract interest. The amount of interest applied could be greater than would be available on a bank account, but still provide the Cab Company with a cheaper form of capital than borrowing from banks. A card containing sufficient capital might provide free transport in that the interest might be sufficient to fund the travel without dipping into the capital.

Purchase of Merchandise on Card

The Travel Card could be used to buy merchandise from the Cab Company other than just journeys. For example, stopping at Service Stations, the customer might buy meals or drinks and other commodities.

Automatic apportionment of Fares

Where systems are operated under licence, and the fare is shared out between local system and network, the fare deducted from a Travel Card would be automatically apportioned between the various interests and credited to the respective bank accounts

Accommodation of Free Transport

Where free travel is provided for senior citizens by the State, the fare would be automatically charged to the Social Welfare Account.

16. Security

Since *Capsules* and *Cab-stops* would operate without drivers and superintending personnel, security would be provided by remote monitoring.

Every Capsule could be equipped with web-cams (cameras) and other sensors to monitor the behaviour of *Passengers*. Where a security risk is observed, a *Capsule* could be prevented from leaving a *Cab-stop*, until the risk is dealt with, or a *Capsule* could be directed to a *Security-Station*. Security personnel could also have a communication channel to *Passengers* in Cabs. (Such details are not necessary to the Invention, but are recommended).

A regulation could be coined to prohibit (or control) the transport of any gun, knife, weapon or metal object (larger than a set of house keys). Sensors could be used that detect metal objects and prevent the entry into a *Capsule* while carrying such items.

Every *Cab-stop* could, likewise, have cameras and/ or other sensors to monitor users seeking entry to the stop. These cameras would be linked to intelligent software capable of recognising threats to the security or safety of the system (e.g., suspicious or excessive goods or luggage) and would notify a security centre which would remotely monitor the users.

On completion of every journey, sensors could be used to detect whether any luggage or other objects is left in a *Capsule*, and prevent departure from the *Cab-stop* until such items are removed. Likewise, any damage to a

Capsule could be spotted and charged, by a notified fine system, to the account of the perpetrator.

If any particular *Route* becomes a security risk, security software could debar it from the *Network* and prevent traffic from the *Route* from accessing the *Network*.

The system would keep a record of all users and journeys for security purposes.

There would be a system for closing *Routes,* emergency access to *Tubes*, redirection of traffic and stopping movement, in the unlikely event of an accident occurring.

17. Allocation of Capsules

Ideally, when a *User* arrives at a *Cab-stop*, there should be at least one *Capsule* waiting for him there.

The *Network* computers would monitor the situation at all *Cab-stops* and dispatch spare *Capsules* from *Cab-Stations* or other *Cab-stops.*

Associated with each *Cab-stop* would be a number indicating the ideal number of *Capsules* that should be found parked at it. The *Network computer* would take *Capsules* from it when overstocked and replenish it when under-stocked. The allocated number would be related to the usual volume of traffic of the *Cab-stop.*

It is probable that many *Cab-stops* would have such low usage that they would not normally have a waiting *Capsule*, but be sufficiently close to a *Cap-Station* to draw, within minutes, from the *Station's* store of *Capsules* when required. Where a user intends travelling from such a Cab-stop (s)he should be able to order a Cab in advance by using a computer App.

18. Goods Delivery

Every customer wishes to have a Cab-Stop outside his own door. It would be technically no problem to have a Goods Delivery Point there as well, adjacent to the Cab-Stop.

While passengers are delivered to a Cab-Stop, goods would be delivered to the Goods-Stop.

Remember, you won't be taking your motor-car with you when you are going shopping. When you have loaded your shopping trolley, you will wheel it to the shop's *Goods-Stop* and dispatch it to your home destination using your Card.

While a typical Passenger Cab would be designed to accommodate up to four passengers and their hand luggage, a typical Goods-Cab would be designed to accommodate a supermarket trolley.

When you return home, your Card would give you access to your goods. You would wheel your trolley home, unpack your trolley, return it to the Goods-Stop and dispatch it back to the store.

Of course, your goods might be a single parcel not requiring a trolley. No problem!

"Click and Collect" will mean "Collect from your own local Goods-Stop."

19. Multi-level Destinations

Krunchie's Cab can be used to deliver travellers to multi-level destinations, as where populations are dispersed through high-rise buildings.

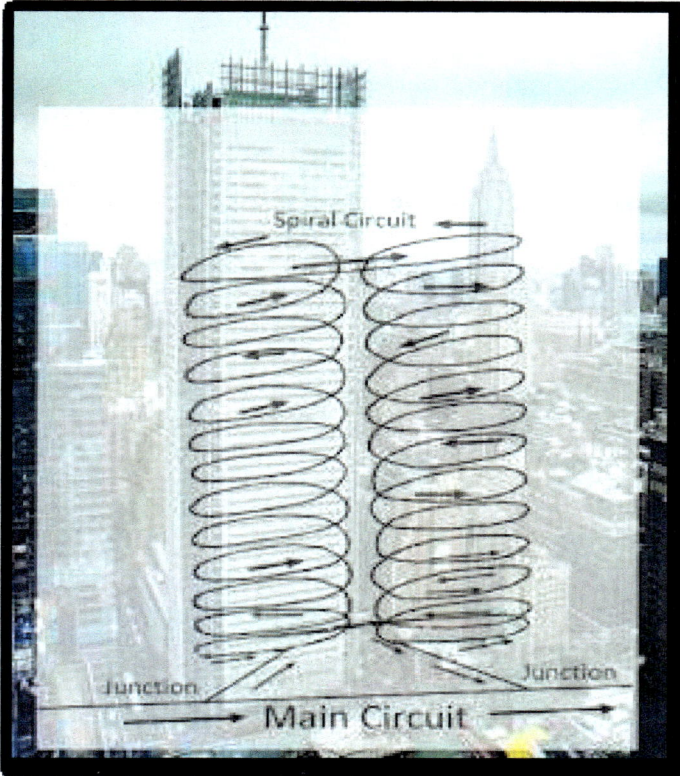

Capsules addressed to a high-rise location depart from the flat Circuit at a Junction that delivers them to a spiral circuit. Rising through the Spiral Circuit, the cabs are propelled, as in any other circuit, by electromagnets. Descending, electromagnets can be used to slow the capsules, feeding back electricity to the system. In effect, the down-travelling cabs help to drive the upward travellers. Cab-stops would be attached to the Spiral Circuit, as to a flat circuit.

20. Electromagnetic Propulsion

As electricity flows through a wire, it creates a magnetic field around the wire. If the wire is formed into a loop, the loop is a magnet, so long as the electricity is flowing. Repeatedly looping the wire to form a coil makes a more powerful magnet.

Since copper can't be magnetised, a coil of copper wire remains a magnet only so long as the current of electricity is flowing. The magnet can, therefore, be switched on an off simply by switching the electric current on and off.

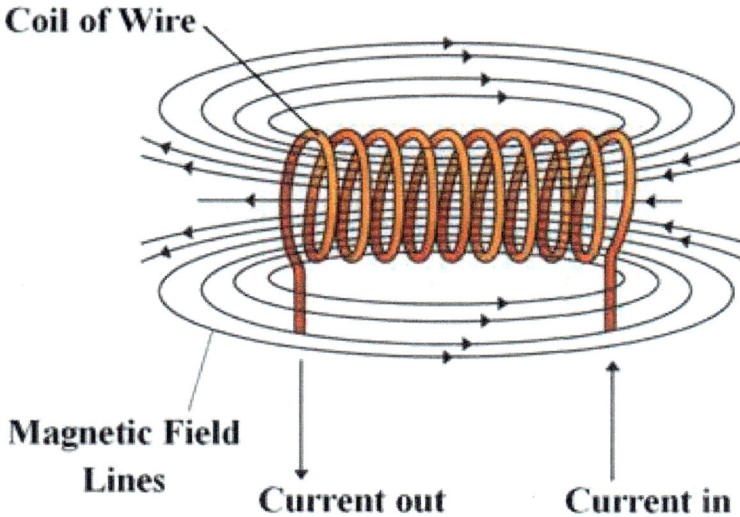

Image from https://standexelectronics.com

The strength of the magnet depends on a number of factors, including the thickness of the wire, the number of loops in the coil and the strength of the current of electricity.

In an electric engine, repeated pulses of electricity are used to propel a piston up and down; and this motion is converted by gears into the turning of a wheel. In *Krunchie's Cab*, the electromagnet is used directly to propel a Cab through a Tube.

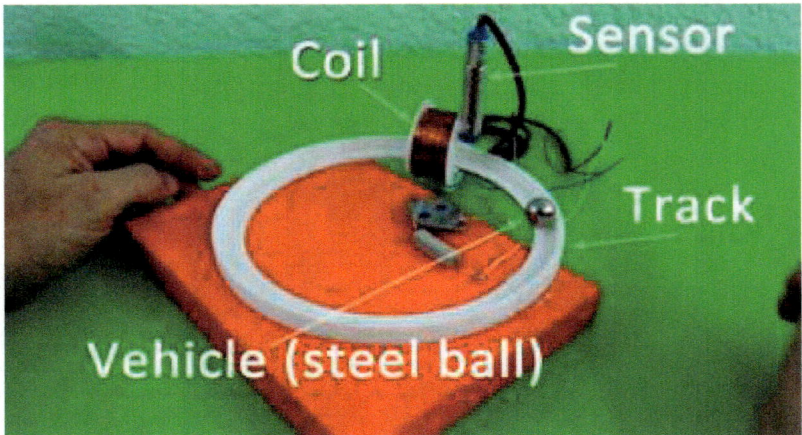

A simple electromagnetic accelerator by Ludic Science on YouTube, shown on YouTube at https://youtu.be/vPjm2FNo5Ok

The vehicle in the picture above is a steel ball that rolls around a plastic track. When the ball comes near the Sensor, it triggers a switch that sends a pulse of electric current through the Coil of copper wire. This turns the Coil momentarily into a Magnet, which boosts the speed of the ball. Thus the ball is kept speeding around the track.

If the magnet were not switched off as the ball passes through, the steel ball, having sped through the loop, would then be attracted back by the magnet, and, instead of continuing around the circuit, would come to rest inside the coil. The accelerator is optimised by switching the current off at precisely the moment where the ball passes through the coil.

In the next picture, from a YouTube video by *Hyperspace Pirate* at https://youtu.be/uNbL3tRZeMQ, a battery-powered circuit with one Copper Coil (right) is compared to another driven by four coils. An astonishingly faster speed is achieved by the latter. This demonstrates that a moving vehicle can be accelerated again and again to achieve high speeds.

The speedier circuit needs a higher outer wall, to prevent the speeding ball from jumping out of the circuit. A Tube would be the most secure way of containing the vehicles within the track, demonstrating an advantage that Tube systems, like *Krunchie's Cab* have over Personal Rapid Transit systems, which have Pods travelling on Tracks.

The desk-top models described in these videos can easily be elaborated in a real-world system. Copper coils (or other electromagnets) can be dispersed around a Circuit. Speed Sensors can be used to monitor the speed of

Vehicles and trip a switch to give them a boost whenever required.

By using multiple electromagnets, dispersed along a Tube, astonishing speeds can be achieved.

The wind created by *Cabs* as they move through a tube would support their movement, making it easier to build up speed on busy routes. The surfaces of Tubes and *Cabs* would be made of modern mutuallly-repelling plastics, which would reduce friction significantly, enabling *Cabs* to glide over the surface of *Tubes* without needing maglev.

Hover-Ball toy, from JML, which has a hard, shiny, base (shown here detached, shiny side down), which enables the ball to glide over any flat surface, even a carpet; no batteries needed for levitation.

Components

While a battery is a device designed to store electric energy, and release it gradually to power devices such as radios, mobile phones or garden tools, a capacitor is a device used to build up a great voltage of electric power and release it in an instant.

Capacitors, photo by Eric Schrader (7189597135.jpg via Wikimedia Creative Commons)

Capacitors can be used with electromagnets to propel pellets or capsules at bullet speed.

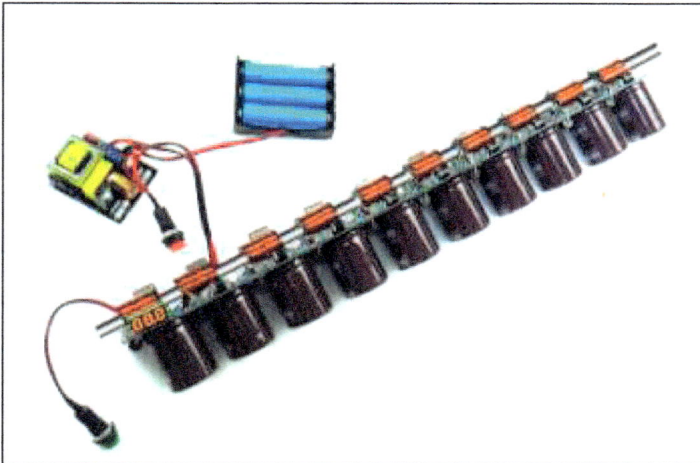

A coil-gun by Tonifishi that can be purchased online (from https://www.AliExpress.com)

Powered by a simple battery-pack, a voltage of about 450 volts is built up in each of the (black coloured) capacitors in the above picture, and released in an instant through the copper coils just as a steel pellet, inside the metal pipe, approaches each coil, accelerating the pellet at each coil and propelling it from the device at bullet-speed.

21. Gravitational Propulsion

Childhood Example

When I was a kid, box-carts were popular. We called them trolleys. We attached wheels salvaged from old prams to planks of wood to make a trolley. Then we stationed our trolley at the top of a slope in the local lane, and, with a gentle push to start, went careering down the slope.

Gravitational propulsion can also be exploited in a modern transport system.

Real-world Potential

The above picture shows a pedestrian bridge (at Castleknock commuter railway station, Dublin). Passengers use this bridge to cross from one platform to the other.

Suppose passengers enter a capsule stationed on the bridge, which then receives a little push sending it down a friction-free slide to commence a journey along the tracks. If the starting point is a mere 26 feet above track level, and the slope of the slide is 45°, by gravity alone the capsule would reach 30 mph by the time it reached track-level, in around 2 seconds (calculated using the gravitational formulae of math-physics and the cosine rule).

Sliding through a tube with minimal friction and air-resistance, a little boost every now and then by electro-magnets is all that is required to keep it going at 30 mph forever.

Gravity can be used, where convenient, in the starting and stopping of Cabs.

Hill and Dale

Gravity can also be used to control the speed of Cabs as they climb and descend hills and dales.

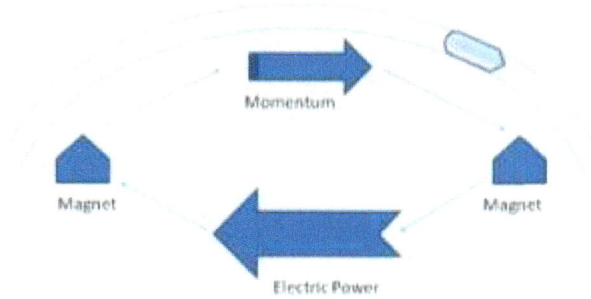

In this diagram, electric power charges an electro-magnet to drive a Capsule up a slope. On the down-slope, magnets are set to resist the momentum caused by gravity, re-generating electric power.

22. Solar Power

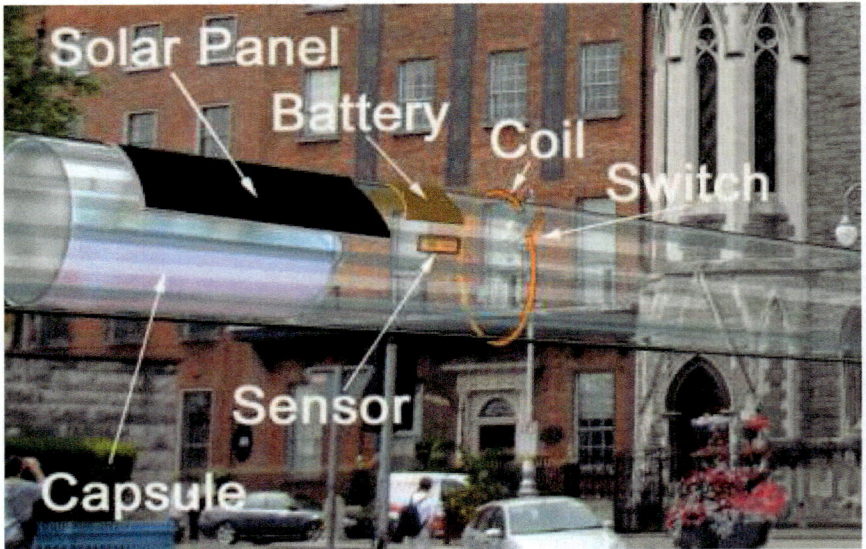

Magnetic Propulsion of a capsule travelling in a tube is provided by an Electromagnet, in the basic form of a Coil of Copper Wire wrapped around the tube.

Optionally, electricity can be generated by a series of Solar Panels attached to the surface of the Tube (or Solar PAINT applied to the entire surface). This Solar Panel would generate electricity from daylight. The electricity generated can be stored in a battery or series of batteries, attached to the Tube surface (and enhanced by capacitors and other electronic components to deliver the required voltage).

In the picture, a Solar Panel charges a Battery, which powers a Sensor, a Switch and an Electromagnet (Coil). The Sensor triggers the Switch, which allows power to flow through the Coil. (The picture shows a single-loop coil, but in actuality a coil would have multiple loops).

A similar arrangement can be used to slow down a Capsule that happens to be travelling above the prescribed speed.

If using Solar Power, a *Krunchie's Cab* system would still be connected to the electricity grid. It would feed the Grid when generating surplus energy and draw from it when short.

The *Tube-Net Transit* system in Hangzhou City, China, is reportedly powered by Solar Panels.

23. Gliders

When we were kids, we made slides, in the snowy weather, by polishing the snow with our feet, until we had a shiny, slippery channel to slide on.

The makers of artificial ski slopes don't go so far. They present a slope that has a degree of grip, so that skiers don't topple over.

Krunchie's Cab will present two surfaces, the surface of the Tube and the surface of the Capsule, which will minimise the friction between the two and maximise the slippiness.

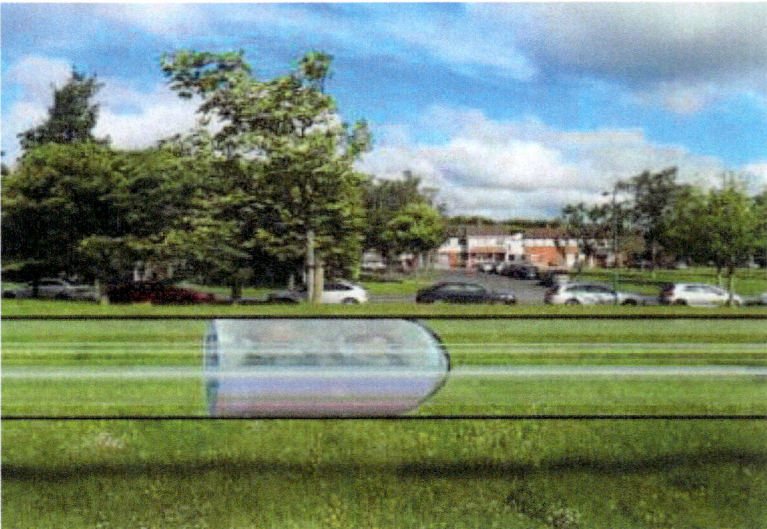

The *Cab* will present with a shiny surface, especially at the bottom, and the Tube will be coated in a material that will repel the plastic of the *Cab.*

The nose of the *Cab* will be curved, so as not to get stuck in the Tube wall, and so as to cut into the air in the Tube, enabling a skinny layer of air to press under the *Cab* to give a slight elevation as the *Cab* glides along. Air friction can be reduced by sucking air from the Tube where there is little movement, but in a busy Tube the air, being pushed along by the *Cabs,* will produce a wind that will assist the movement.

24. Tube Laying

Easy Lay

Tubes can be laid anywhere, on hard or soft land or under water, with minimal site preparation.

Picture from a *Plastics Pipe Institute* brochure: www.plasticpipe.org

Tube Cross-overs

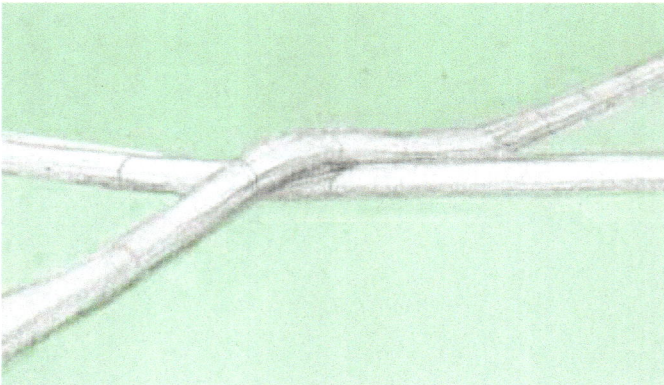

Where routes cross over each other, there is no intersection. One Tube is simply raised above the other.

Similarly, where a route crosses over a road, railway or other phenomenon, the Tube is raised above or constructed under the obstruction.

Under Sea

Existing cable-laying techniques and patents for under-sea tube-systems can be utilised in constructing the under-sea connections. Laying 5-foot tubes under-sea is a low-cost operation compared to constructing under-sea road tunnels.

Cheaper than Roads or Railways

Tubes are much cheaper to install than roads or railways.

Tubes can be laid over any surface, soft or hard; can be laid under or over water, in tunnels or over bridges, and can easily be raised on stilts where required.

Accommodates more Traffic

A tube for *Krunchie's Cab* takes up no more space than a footpath.

One tube can accommodate more motor-traffic than four road lanes.

Can Use existing channels

Tubes can be laid beside existing roads, railways and canals, and along disused railway lines and through existing tunnels. A network can be quickly constructed by using these existing channels.

Charlestown – Swinford (Ireland) disused railway (begging for a *Krunchie's Cab* circuit

Elevation

Elevating Tubes for *Krunchie's Cab* is much simpler and cheaper than road-elevation, and so quick to implement!

Substantial structures used to elevate roads above other channels or obstructions

Light Structure used to elevate Krunchie's tube above a canal or railway or road (artist's impression).

25. Construction and Materials

Prefabrication

Tubes would be prefabricated in different lengths, curvatures and specifications, and transported for assembly on site.

Various Specifications

There would be several Section specifications, depending on the amount of curvature, flexibility and stress-bearing required in different parts of a Route, and particular functions served, for example, at cab-stops. A Route would be constructed pursuant to an Engineer's instructions specifying the exact type and curvature of each part.

Variation per Terrain

Tubes can be laid on different surfaces: hard ground, soft ground, marshy ground, under water, elevated on stilts, over roof-tops, and so on. There should be a certain amount of flexibility, to allow Tubes to bend to the terrain over which they travel, but higher-speed Routes would be designed with less curvature than lower-speed Routes. Where Tubes cross each other, one Tube might simply be laid resting on the other, or, alternatively, supports used to achieve the elevation, as found convenient by the construction engineer, subject to specifications laid down in the licence.

Materials

The materials used in the construction of Sections of Tubes would vary in accordance with the stress to be borne, insulation desired and flexibility to be allowed. Lighter, flexible, material would be possible in low-speed Routes.

Stronger material would be necessary in Sections bearing the weight of crossing Tubes, or Tubes bridging valleys, or laid under lake or sea, etc.

Opaque or Transparent

Opaque materials would be suitable for most Tubes. However, for some Routes in scenic areas, transparent Tubes would be appropriate. Likewise, in urban areas, transparent Tubes might often be advisable to minimise the visual obtrusiveness of the Tubes. Alternatively, Tubes could be covered and/or built into the design of roads to maximise pedestrian convenience and visual attractiveness.

Since it is envisaged that some Tubes would be made of transparent material, the upper part of Capsules should, also, be transparent.

Insulation of Tubes

In icy regions, tubes could be encased in insulating material.

Rapid Implementation

Existing road and rail links could be used for rapid implementation. A Tube would take up a fraction of the space of a traffic lane in a road system and could be installed along the side of existing roads, with little reduction in existing road capacity, or elevated on stilts so that the road-capacity is not reduced at all. Many railways will be found to have sufficient space to accommodate a Capsule Tube beside the existing track.

Locations

New Tubes can be laid across virgin territory with little or no site preparation.

Tubes can be laid on existing roads and railway lines.

Tubes can be elevated above houses – and Cab-stops provided on roof-tops.

Many underground passages built for the purpose of accommodating metro lines or sewers will be found to be suitable for use by Tubes.

Tubes can cross over existing junctions or other obstacles by being raised on stilts.

26. Skeleton of World-wide Network

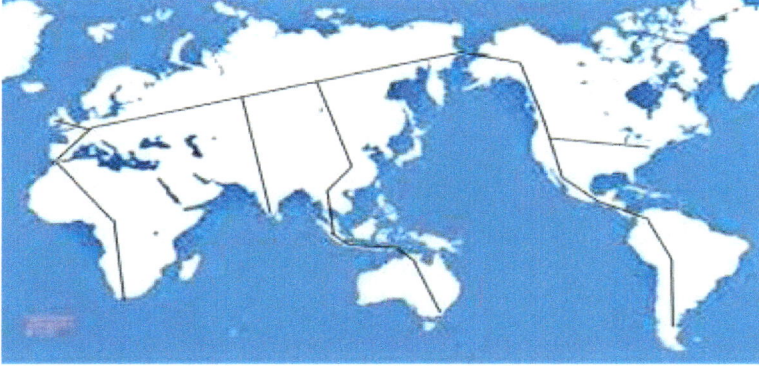

This illustration shows a skeleton of a possible World-wide network, which minimises sea-crossings. The entire land-mass of the earth can be connected, excluding remote islands whose population-density might not justify the expense of a sea-connection.

Existing cable-laying techniques and patents for under-sea tube-systems can be utilised in constructing the under-sea connections.

The longest route on the map is the route from Cape-town in South Africa to Buenos Aires in South America. Using best ET3 technology and Maglev, that trip would take only 3 hours.

At present, though the air-distance is not as long as the land-distance, the best travelling time available between these two destinations is 9 hours[1]. Add to this the time spent travelling to and from airports and waiting at airports, and constrictions imposed by schedules.

---

[1] Per Travelmath (http://www.travelmath.com/) (31 March 2014): "The flight time from Cape Town, South Africa to Buenos Aires, Argentina is: 9 hours, 3 minutes."

27. Route Capacity

Capacity of Local Tube: 500,000 persons per day

If a Capsule, 6 feet in length, travelled through a tube at 30 miles per hour (the speed of our local routes), then the following calculation shows that a single Tube could carry up to 500,000 passengers per day, allowing for a gap of 24 feet between Capsules (true for Routes up to 30 miles long; most local Routes would be shorter, and longer Routes would be faster).

30 mph = 158,400 ft per hr (30 X 5,280)

6 ft capsule + 24 ft gap = 30 ft per capsule

→ Max No. of Capsules passing a point per hour:

158,400 ÷ 30 = 5280 per hr

Average 2 passengers per capsule → 10,560 persons per hour

At least 2 journeys per Route (into and out of town)

→ Capsule Route capacity ≥ 21,120 passengers per hour

= 506,880 per day (21,120 X 24)

Existing Dublin Bus Route: 3,000 passengers per day

According to Irish Minister for Transport, Martin Cullen, speaking in 2005, Bus Routes in Dublin carry an average of 3,000 passengers per day. Replacing a Bus Route by a Tube Route would vastly increase the Capacity and Potential

Income of the Route, (while significantly reducing operating cost).

Multiple Routes: savings multiplied

While it would be very advantageous to replace a single bus route by a tube route, the advantages are multiplied a thousand times when multiple routes are inter-connected to effect door-to-door transport.

Faster Routes: greater capacity

Speed of travel in Ring Routes would be twice that of local routes, and Inter-city routes would be twice that again. The faster the route the greater the capacity!

Joining Capsules to form Cab-trains and Clustering of Tubes

The capacity of routes can also be increased by joining Capsules together temporarily using a magnetic or physical hook, to form a Capsule-train, or, for super-busy routes, clustering several tubes together, just as electric wires are clustered to form a cable.

The speeds claimed here for the system are very modest.

28. Why previous Tube Systems failed

Point to Point: Motor cars provide door-to-door transit. Previous Passenger Tube Systems, like trains, offered station-to-station transit only.

Pneumatic Propulsion: Pneumatic propulsion was proposed for Tube Travel. Diesel and electric trains, on rails, won the day for underground systems. Diesel or electric engines could be used in *Krunchie's Cab*, but I reckon that electro-magnetic propulsion would be better.

Stop and Start: in previous systems, when one vehicle travelling in a Tube (or on a train-track) stopped, so did all the other vehicles on the track. In my system, stopping and starting are off-line: there is continuous movement, at more-or-less constant speed, without stopping, within the Tubes. The stopping of a Cab does not hold up other Cabs.

Personal Rapid Transit and Urban Light Transit: These are less flexible, more expensive, and have not developed the interconnecting and globalising features.

29. No More Waiting or Queuing

Krunchie's Cab would be available on demand at all times of day and night. No need to wait for the scheduled arrival!

Bus and train services are slowed down by bus-stops and train-stations, whenever the vehicle is required to stop. There is no stopping on Krunchie's Cab, except at destinations.

Many more Cab-stops can be provided along a Krunchie's Cab route than Bus-stops, since additional cab-stops do not mean additional delay. For example, at the date of writing this, there are 80 Bus-stops on Dublin's Number 40 Bus-route, spaced an average of about 350 yards apart. Cab-stops could usefully be provided every 100 yards.

When a customer arrives at a Cab-stop, (s)he will normally find a Capsule waiting. If not, a Capsule would be dispatched to arrive within a minute.

Railway stations are necessarily spaced widely apart, and intermediate stops are often by-passed in order to provide express service. Krunchie's Cab would provide Cab-stops wherever there are people to be served, and service would be available always at all stops and would always be express.

30. How Krunchie's Cab beats Motor Cars

No driving skill needed.

No exhausting driving: users can relax completely while travelling.

No capital investment, driving licence, insurance fee, or service cost.

No danger of crashing.

No parking problems.

No hold-ups.

No wrong turns or running out of fuel.

Reliable delivery to one's destination: all the user must do is show travel card (or credit card or mobile phone) to the system and key in his destination.

Less expense: tube travel is cheaper than motor travel.

Motor Cars are obsolescent, due to increasing world population and resulting traffic congestion. Construction of roads can't keep up with increasing demand for travel. Tube is the travel of the future.

In a Tube system, there are no hold-ups, no parking problems, no exhausting driving, no accidents, and the destination is reached much faster.

31. How Krunchie's Cab beats Buses

You normally wait for a bus. In Krunchie's system, a Cab is normally waiting for you.

Only one of multiple buses is yours. You wait at your bus-stop while several other buses pass, before yours comes along. In Krunchie's system, you take the first vehicle that comes along, since every single capsule can take you to your destination.

You might need 2 or more buses to get to your destination, but only one Cab.

The bus will rarely take you door-to-door. Krunchie's Cab always will.

Buses are more expensive: they use more fuel and require more personnel and maintenance.

One Tube route can handle a hundred times more passengers than a similar Bus route.

A Tube takes up less road space and can be raised off the ground, if required.

Buses cause and are affected by traffic congestion: Tubes are congestion-free.

Buses are subject to road accidents. Tubes are accident-free.

Bus services are subject to strikes. *Krunchie's Cab* is strike free.

32. How Krunchie's Cab beats Trams and Metro

Tubes are easily laid compared to rail systems.

Tubes can go where train tracks can't: up hills and down dales, on hard or soft land, raised on stilts over roads, railways, houses or other obstructions, and under the sea.

A cab-stop can soon be provided near everybody's front door.

There is only one cab needed to take you to any destination.

No need to plan your journey: routing is automatic.

No schedule to meet: travel at your choice of time.

No bus or taxi journey is needed to reach your cab "station." The stop's outside your door!

Less construction work;

Routes are more flexible;

New routes can be added continuously and existing routes expanded.

Faster expansion of the route system;

Cheaper vehicles;

Greater route capacity;

Greater speed; no intermediate stops;

Krunchie's Cab

Greater gradient tolerance (i.e., ability to go up and down hills);

More privacy and comfort.

33. How Krunchie's Cab beats Hub travel

There is a worldwide trend towards the creation of massive hubs where travellers can transfer from one route and/ or mode of transit to another.

Krunchie's Cab does not drop you off at a massive hub. It brings you all the way home, or to wherever you want to go!

34. How Krunchie's Cab beats Hyperloop and ET3

Hyperloop, or ET3 ("Evacuated Tube Transport Technology"), will bring you at great speed from one large centre of population to another.

"Speed in initial ET3 systems is 600km/h (370 mph) for in-state trips, and will be developed to 6,500 km/h (4,000 mph) for international travel that will allow passenger or cargo travel from New York to Beijing in 2 hours" (from www.et3.com).

While Krunchie's Cab is merely a concept, *Hyperloop* has attracted millions of dollars of investment and engaged multiple teams of engineers and designers and a large body of lobbyists. However:

Hyperloop will not pick you up at your doorstep and deliver you to the doorstep of your destination. First you must find a Hub from which to start your journey. When you get to your *HYPERLOOP* terminus, you still have to find your way to your local destination.

HYPERLOOP engages complex and expensive technologies. Krunchie's Cab uses simple and easily-implemented technology.

Krunchie's Cab can be tried in a pilot scheme on any few miles of disused railway or along an existing road.

Krunchie's Cab

HYPERLOOP requires a large distance; San Francisco to New York has been proposed.

Krunchie's Cabs can get off to a quick start by following existing road and rail routes. Super-high speed systems require very straight routes. Providing such routes has legal, administrative, engineering and cost implications and takes time to implement.

While *HYPERLOOP* has years of a head-start, Krunchie's Cab can still be up and running first.

Even without *HYPERLOOP*, Krunchie's Capsule offers international and trans-continental travel at speeds that are very competitive with existing modes of transport.

35. How Krunchie's Cab beats Airplanes

No waiting at airports;

No travel-planning between airports and local destinations;

No schedule times to meet;

Virtually no capacity limit;

No long or short term parking fees;

No cancellations due to weather or industrial disputes;

Less risk of accident;

Ultimately, with connection to super-high speed tube systems, shorter travelling time.

36. How Krunchie's Cab beats the Self-drive Car

The self-drive car does not solve the principal problems of the motorcar, viz., gridlock and unsustainable roads (but Krunchie's Cab does)!

The self-drive car is extremely complex, with a great number of cameras and sensors and miles of software-code. Krunchie's Cab is a very simple device - far simpler, indeed, than even the fossil-fuel car.

Nobody knows yet if the self-drive car will ever be a success. It is impossible to be sure that it can foresee all obstacles in the physical environment. It is absolutely undeniable that Krunchie's Cab will actually work; it merely requires providing tubes and matching capsules and a simple operating protocol.

The self-drive car requires a large team of highly-paid experts at the cutting edge of technology. Krunchie's Cab requires a small team of competent engineers.

The self-drive car will only take you on roads. Krunchie's Cab will take you anywhere and everywhere.

Compared to the self-drive car, Krunchie's Cab is low-tech, simple to manufacture, cheap, reliable, and potentially universal in scope.

Pedestrians and wild-life are not endangered by Krunchie's Cab, as they are by motor transport, including the self-drive car.

37. Advantages of Krunchie's Cab over other forms of transport:

Door-to-door transport;

No need for users to plan routes or switch vehicles or modes of transport;

No stopping and starting en route;

No driver needed;

Schedule-free: constant availability around the clock;

Enhanced access to destinations;

No traffic jams;

Reliable calculation of travel time;

Potential world-wide connectivity;

Accident free;

No traffic violations – reduced workload of courts;

Low cost of operation;

Low cost of construction compared to roads and rail;

Few engineering problems compared to road construction;

Flexibility of implementation compared to rail;

Pre-fabricated parts and minimal site-preparation means Rapid Construction;

Rapid implementation by using existing road and rail routes;

Eliminates or reduces the need for new roads and railways;

Reaches all communities no matter how remote; revives rural communities;

Re-unites communities divided by motorways or railroads;

Sustainable, self-generating energy usage;

No carbon emission, due to magnetic propulsion fuelled by solar panels;

Redresses global warming by reducing energy consumption and reliance on fossil-fuel;

Virtually silent operation with magnetic propulsion;

Quiet streets – eliminates or substantially reduces motor traffic;

Returns roads to pedestrian use;

No danger to wild life, cyclists or pedestrians;

Significant wealth released back from roads to human well-being.

38. Implementing Krunchie's Cab

Implement in a single Route

Krunchie's Cab can be implemented in a single route. It will be more efficient than a bus route, take up less road space, deliver passengers more speedily to their destinations, use less energy, and accommodate a much greater number of passengers.

Pilot Route on disused Railway

A pilot route might be implemented on a few miles of a disused railway, or in a park as a tourist attraction.

Urban Ring Route an Early Implementation

An early implementation might be a ring route that would circle an urban area, connecting to multiple bus and rail routes.

Join separate Routes to form Network

When several individual routes are implemented, they can be joined, and automatic routing provided across the network.

Networks can be expanded organically and inter-city links established.

First to Market can lay down Standards

The first to market with Krunchie's Cab could licence their brand world-wide, with specifications for every element of the system, thus producing a consistent international network.

International routes and trans-continental routes would follow.

Super-fast Tubes

Long-distance Circuits can be enhanced by technologies of air blow and suction to create super-fast speeds to compete with *Hyperloops*.

39. Summary

Krunchie's Cab is a proposed system of travel where motor-car sized Capsules, travelling in Tubes, provide Door-to-Door transit.

Travel is non-stop between starting-point and destination, and routing is automated.

Low-speed (30 mph) local routes are connected to medium-speed Ring Routes and high-speed Inter-City Routes and International Routes.

No venue, no matter how remote, needs to be excluded from the system (except on remote islands where an under-sea route might not be viable).

Krunchie's Cab beats and can replace motor-cars, buses, trains, trams and airplanes.

Existing technology in pipe-manufacture, vehicle manufacture and control systems for automatic routing can be engaged to bring a system into quick production.

The system can begin as a single route and be developed organically and by licensing the technology worldwide.

Krunchie's Cab

Published by the author 2023

(Formerly published under the title "Transport 21 Hundred" (2017)

ISBN: 9798372385061

Imprint: Independently published

Author's email address: KrunchieKilleen@gmail.com

Web Site:

https://krunchiescab.blogspot.com/

Available in paperback from Amazon

Available on Kindle

Printed in Great Britain
by Amazon

27713396R00056